WHALES

By Jane Werner Watson

Illustrated By Rod Ruth

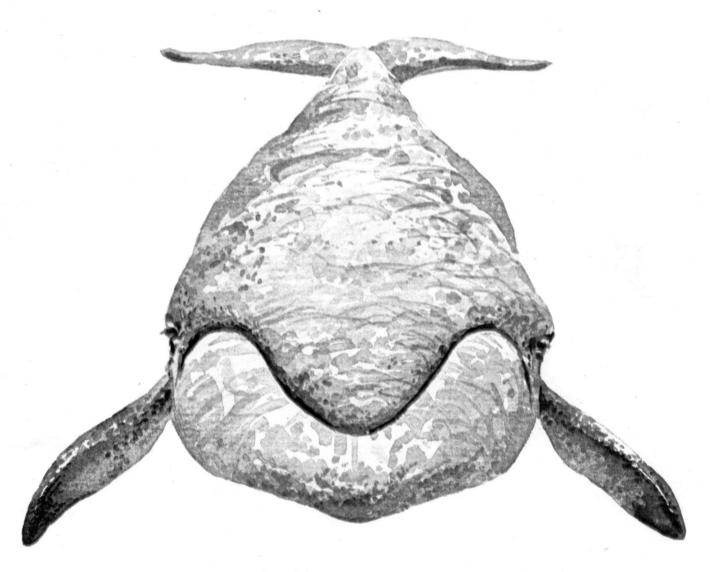

GOLDEN PRESS
Western Publishing Company, Inc.
Racine, Wisconsin

© 1978 by Western Publishing Company, Inc.
All rights reserved. Produced in U.S.A.

GOLDEN®, A GOLDEN BOOK®, and GOLDEN PRESS®
are trademarks of Western Publishing Company, Inc.
No part of this book may be reproduced or copied in
any form without written permission from the publisher.

0-307-10824-4

Library of Congress Cataloging in Publication Data

Watson, Jane Werner, 1915-
  Whales.

  SUMMARY: A Brief introduction to the members of the whale family including the blue whale, narwhal, and dolphin.
  1. Cetacea—Juvenile literature. [1. Whales]
I. Ruth, Rod. II. Title.
QL737.C4W33 1979   599'.5   79-10589

Have you ever seen a fountain spouting out at sea? If you have, it was a whale blowing out its breath.

Whales can swim deep in the ocean, but they have to come up to the surface for air.

Fish don't have to surface to breathe. However, whales are *not* fish, though they do live in the sea and even look like fish. Whales are animals called *mammals*.

All mammals must breathe air into their lungs. Whales breathe through blowholes on top of their heads instead of through noses, as we do.

Fish lay eggs that hatch into baby fish, but whales, being mammals, give birth to their babies.

A baby whale is born tailfirst, so its tail gets used to moving before its head appears.

This is good, because a newborn whale has to be able to swim right away. It must swim to the surface to take its first breath of air.

Mother whale helps, of course.

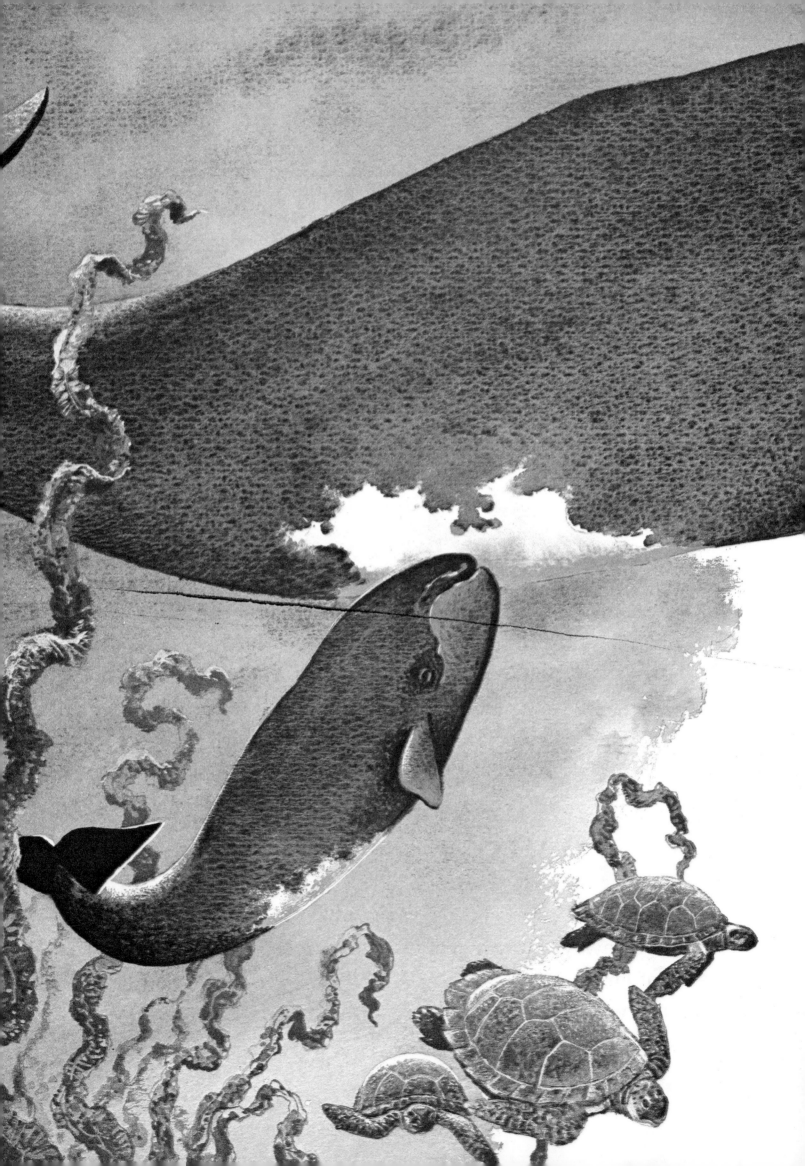

Mother whale gives her baby loving care. She guides it with her flipper. She squirts her milk into its mouth until the baby is ready to eat solid food.

Most mammals have legs and some hair.

The whale family of mammals moved from the land into the sea many millions of years ago.

After a long time, their front legs turned into flippers. The whales also lost their hair. Smooth, hairless skin is much better for swimming.

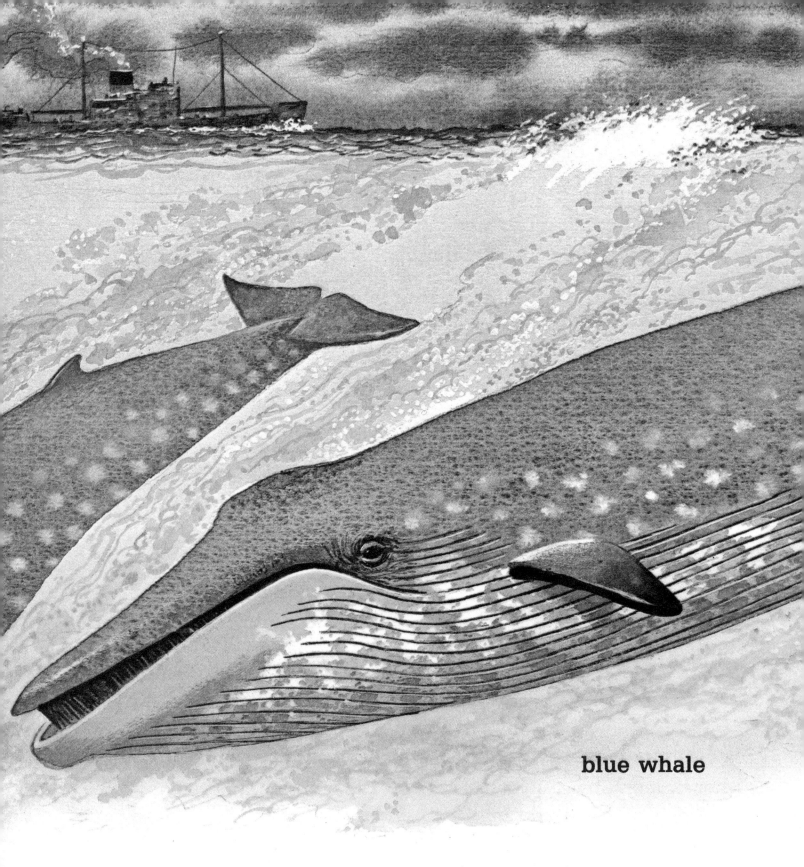

blue whale

Gradually, whales' bodies became streamlined, with big heads and no necks at all.

Whales became splendid giants of the sea!

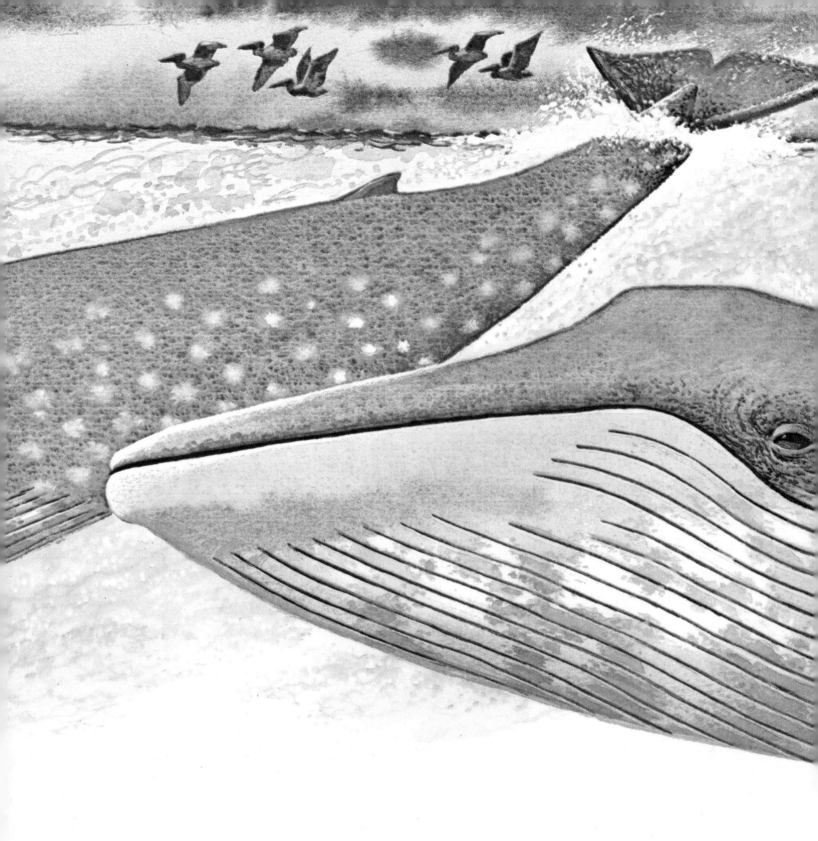

Some whales are the biggest animals alive.

Biggest of all are the blue whales. They may be one hundred feet long! Blue whales are bigger than the largest dinosaur that ever lived!

Blues are baleen, or "moustache," whales.
These whales wear their moustaches *inside* their mouths.
A whale's moustache is not made of hair. It is made of long, tough strands called whalebone, or baleen.

The whale swims with its mouth wide open, and seawater flows in. In the seawater are very small plants and animals called *plankton.*

When it has taken in a good lot of water and plankton, the whale closes its mouth. Then it squirts out the seawater between its lips.

But its moustache of baleen traps the plankton inside, and the whale swallows this tasty meal.

Not all whales have moustaches of baleen. Some whales have teeth instead.

The largest with teeth is called the sperm whale. It may grow to be sixty feet long.

Toothed whales catch fish with their cone-shaped teeth. Some also like to eat big ten-armed squid.

sperm whale

killer whale

Killer whales are very fierce hunters. They even eat walrus and seals.

They only hunt when they are hungry, however. Killer whales who are kept in aquariums, where they are fed every day, become very friendly.

They can learn to do tricks, and they will even let swimmers ride on their backs.

Most whales have gray skins. But a few toothed whales in the far north turn almost white as they grow up.

The bottle-nosed whale is yellowish, and the beluga whale's skin is white.

bottle-nosed whale

The beluga whale is very sociable and likes to swim in large groups. Many other whales do this, too.

The beluga is famous because it makes such musical sounds. Sailors have called it "the sea canary."

beluga whale

narwhal

The narwhal is a very unusual whale.
It has one upper tooth that grows to be six to ten feet long! It is like a slender, ivory horn.

Smallest of the whales are the dolphins and the porpoises. Some are so small that they can live in shallow, muddy rivers.

dolphin

Most porpoises have blunt noses, and most dolphins have sharp beaks. Otherwise, they are very much alike.

Like most whales, dolphins and porpoises are very intelligent. They can solve problems. They even have a language of sounds.

They like to play games and perform tricks. They like to leap high out of the water as gracefully as dancers.

They are good-natured and are friendly to people and to one another.

No wonder they are the brightest stars in many water shows, these wise and wonderful whales.